YOUR KNOWLEDGE HAS VALUE

- We will publish your bachelor's and master's thesis, essays and papers

- Your own eBook and book - sold worldwide in all relevant shops

- Earn money with each sale

Upload your text at www.GRIN.com
and publish for free

Bibliographic information published by the German National Library:

The German National Library lists this publication in the National Bibliography; detailed bibliographic data are available on the Internet at http://dnb.dnb.de .

This book is copyright material and must not be copied, reproduced, transferred, distributed, leased, licensed or publicly performed or used in any way except as specifically permitted in writing by the publishers, as allowed under the terms and conditions under which it was purchased or as strictly permitted by applicable copyright law. Any unauthorized distribution or use of this text may be a direct infringement of the author s and publisher s rights and those responsible may be liable in law accordingly.

Imprint:

Copyright © 2014 GRIN Verlag, Open Publishing GmbH
Print and binding: Books on Demand GmbH, Norderstedt Germany
ISBN: 9783668555020

This book at GRIN:

http://www.grin.com/en/e-book/378123/the-technology-acceptance-model-tam-an-overview

Johannes Köck

The Technology Acceptance Model (TAM). An Overview

GRIN Publishing

GRIN - Your knowledge has value

Since its foundation in 1998, GRIN has specialized in publishing academic texts by students, college teachers and other academics as e-book and printed book. The website www.grin.com is an ideal platform for presenting term papers, final papers, scientific essays, dissertations and specialist books.

Visit us on the internet:

http://www.grin.com/

http://www.facebook.com/grincom

http://www.twitter.com/grin_com

Name:	Johannes Köck

Assignment:	Assignment #1 - Applying a theoretical lens

1. About the theory:

The Technology Acceptance Model (TAM) is an information systems theory. This model was developed by Fred Davis in his dissertation which was published in 1989. Since then, this model has spread to one of the most cited models in the context of technology diffusion (Kotrík). User acceptance of technology has been a vital area of studies for two decades now. Many models do predict the diffusion of a system but the Technology Acceptance model is the only model which focuses mainly on Information Systems (Chuttur).

With a growing demand for technology in the 1970's the increasing failure of adapting systems within enterprises became a new area of research. Fred Davis, a doctoral student at the MIT Sloan School of Management, proposed the Technology acceptance model in 1985. He explained that the use of a system is a response to user's motivation. User's motivation on the other hand depends on system features and capabilities. (Chuttur)

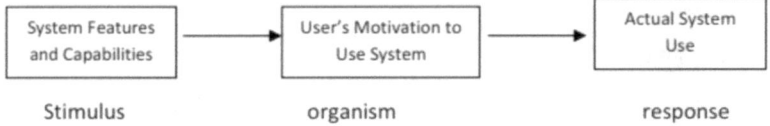

Figure 1: Conceptual model for technology acceptance (Davis, 1985, S. 10)

Davis had two goals with his new model: (Davis, 1985, S. 2)

- To improve the understanding of the processes of technology acceptance and thereby gain theoretical insights for the design and implementation of Information Systems
- "TAM should provide the theoretical basis for a practical user acceptance testing methodology that would enable system designers and implementers to evaluate proposed new systems prior to their implementation"

In August 1989 the Technology Acceptance Model was published in an article by Davis, Bagozzi and Warshaw "User Acceptance of Computer Technology: A Comparison of two Theoretical Models" (Management Science 35(8), 982–1003). In this article a slightly adapted TAM is proposed. According to google scholar this articel has been cited approximately 11.500 times up to now. In September 1989 Davis published his article „Perceived Usefulness, Perceived Ease of Use, and User Acceptance of Technology". In this article Davis explains basic elements of TAM (Kotrík). This journal paper has been cited 21.300 times up to now.

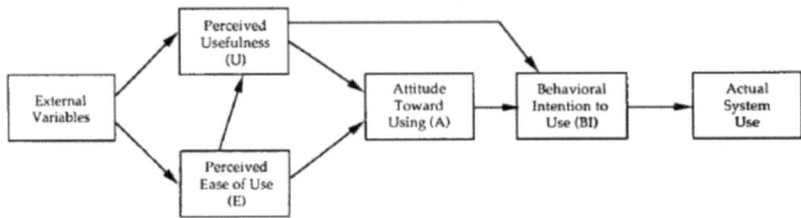

Figure 2: Technology Acceptance Model (Davis, Bagozzi, & Warshaw, 1989, S. 985)

The Technology Acceptance Model is an adaption of TRA (Theory of Reasoned Action, Ajzen and Fishbein 1980), a model from social psychology. However, Davis states, that the *perceived usefulness* and *perceived ease of use* are of "primary relevance for computer acceptance behaviors" (Davis, 1989, S. 985). Perceived usefulness (U) is defined as "prospective user's subjective probability that using a specific application system will increase his or her job performance within an organizational context". Perceived ease of use (E) is the "degree to which the prospective user expects the target system to be free of effort". (Davis, 1989, S. 985) Davis states further, that computer usage is determined by Behavioral intention to use (BI). And the other way around, BI is determined by the "person's Attitude toward using a system (A) and perceived usefulness (U)". (Davis, 1989, S. 985)

$$BI = A + U \qquad (1)$$

In the following the different elements of TAM will be explained. Davis hypothesized, that Attitude toward using (A) and perceived usefulness (U) are the major determinants of whether a system will be used or not. The U-BI relationship in TAM represents the "direct effect" that people have intentions with regard to using a system largely based on their beliefs of how a system will increase their performance (usefulness) (Davis, 1989, S. 986). Hence, U has a direct effect on BI over and above A. However, the attitude toward using (A) is determined by perceived usefulness (U) and perceived ease of use (E):

$$A = U + E \qquad (2)$$

According to TAM, U has a positive influence on A and E has important effect on A as well. There are mechanisms by which E influences Attitude (A). "The easier a system is to interact with, the greater should be the user's sense of efficacy ... regarding his or her ability to carry out ... to operate the system" (Davis, 1989, S. 987). Improvements in E may also contribute to increase performance. Consequently, a person might be able to "accomplish more work for the same effort". (Davis, 1989, S. 987) Nevertheless, E directly effects on usefulness (U):

$$U = E + External\ Variables \qquad (3)$$

Perceived usefulness (U) and the Perceived ease of use (E) are distinct but related elements. The Perceived usefulness (U) can be influenced by external variables without considering the ease of use (E). For instance, looking at two different forecasting systems which are equally easy to use (E). One system will provide more accurate forecast and it will be recognized as the more useful system (U) with E parity at the same time (Davis, 1989, S. 987). Or if one system provide better visualization of data this system will be seen as the more useful system. So the characteristics of a system directly influence U besides the indirect effects of E on U. (Davis, 1989, S. 987) Perceived ease of use (E) is determined by external variables:

$$E = External\ Variables \qquad (4)$$

Many of the features such as menus, icons, mice and touch screens are specifically developed to improve usability. Training, documentation and user support are other examples of external variables in order to positively influence the ease of use (E) (Bewley, Roberts, Schiot, & Verplank). After introducing the elements of TAM in this section, the following section should take a closer look on how this model is applied in research.

2. Theory in IS literature

Looking for TAM in current research leads to a variety of TAM applications in research. Random examples of published research paper in this area in well-known journals are for instance:

- "Understanding the acceptance of teleconferencing systems among employees: An extension of the technology acceptance model" (Park, Rhoads, Hou, & Lee, 2014)
- "Applying the Technology Acceptance Model to the introduction of healthcare information systems" (Pai & Huang, 2011)
- "Investigating User Resistance to Information Systems Implementation: A Status Quo Bias Perspective" (Kim & Kankanhalli, 2009).

Hence, it seems that TAM is still a topic of interest in science. The mentioned study about the acceptance of teleconferencing systems in organizations is one up to date example for the application of the TAM framework on current research areas. This study aimed to understand how employees use these teleconferencing systems for work-related meetings. The dominating factors for the use of these systems have been examined. This study came out with personal and organizational factors directly influencing teleconferencing systems use. Basically, this study confirmed the key propositions of TAM and displayed that individual factors including anxiety and self-efficacy and organizational factors like support are the external variables. These variables directly affect perceived ease of use (E) and perceived usefulness (U) of teleconferencing systems. (Park, Rhoads, Hou, & Lee, 2014) In order to support the further acceptance of respective systems (teleconferencing systems) these variables (anxiety, self-efficacy, institutional support, voluntariness) have to be taken into focus. With specific evolvement of influencing variables the actual system's use can be increased.

3. What insight can IT management gain form TAM?

IT management can gain crucial information from TAM. As IT areas within an enterprise have to do with the implementation of new technologies TAM could support IT prior to the implementation of new technologies. Once a technology is in the "innovators' phase" measurement scales for this technology could be developed. These scales (example see appendix) are in the given context of the new technology. Respondents might be employees which are supposed to use this new technology in future. Responses from respondents can be analyzed and furthermore external key variables should be figured out. These key variables, indirectly affecting actual systems use, should be in focus of IT management in order to get a http://waltoncollege.uark.edu/images/faculty/FDavis.jpgnew technology from the "innovators' phase" to a phase where technology is accepted by the majority of respective staff.

Bibliography

Bewley, W., Roberts, T., Schiot, D., & Verplank, W. (December 1983). Human Factors in Computing Systems. *ACM*, S. 72-77.

Chuttur, M., *Overview of the Technology Acceptance Model: Origins, Developments and Future Directions.*

Davis, F. (1985). *A technology acceptance model for empirically testing new end-user information systems - theory and results.* Massachusetts.

Davis, F. (1989). Perceived usefulness, perceived ease of use, and user acceptance of information technology. *MIS Quarterly*, 319-340.

Davis, F., Bagozzi, R., & Warshaw, P. (August 1989). User acceptance of computer technology - a comparison of two theoretical models. *Management Science*, S. 982-1003.

Kim, e.-W., & Kankanhalli, A. (2009). Investigating User Resistance to Information Systems Implementation: A Status Quo Bias Perspective. *MIS Quaterly*, S. 567-582.

Kotrík, J., *UTAUT - die gegenwärtige Weiterentwicklung von TAM.*. Wirtschaftsuniversität Wien.

Pai, F.-Y., & Huang, K.-I. (May 2011). Applying the Technology Acceptance Model to the introduction of healthcare information systems. *Technological Forecasting and Social Change*, S. 650-660.

Park, N., Rhoads, M., Hou, J., & Lee, K. M. (October 2014). Understanding the acceptance of teleconferencing systems among employees: An extension of the technology acceptance model. *Computers in Human Behavior*, S. 118-127.

Appendix

1. 14 statements towards the use of an electronic mail system (rating from 1 to 7 on a like scale):

Initial scale items for perceived usefulness (Davis, 1989, p. 324)

Item No.	Candidate item for measuring for perceived usefulness
1	My job would be difficult to perform without electronic mail.
2	Using electronic mail gives me greater control over my work.
3	Using electronic mail improves my job performance.
4	The electronic mail system addresses my job-related needs.
5	Using electronic mail saves me time.
6	Electronic mail enables me to accomplish tasks more quickly.
7	Electronic mail supports critical aspects of my job.
8	Using electronic mail allows me to accomplish more work than would otherwise be possible.
9	Using electronic mail reduces the time I spend on unproductive activities.
10	Using electronic mail enhances my effectiveness on the job.
11	Using electronic mail improves the quality of the work I do.
12	Using electronic mail increases my productivity.
13	Using electronic mail makes it easier to do my job.
14	Overall, I find the electronic mail system useful in my job.

Initial scale items for perceived usefulness (Davis, 1989, p. 324)

Item No.	Candidate item for measuring for perceived usefulness
1	My job would be difficult to perform without electronic mail.
2	Using electronic mail gives me greater control over my work.
3	Using electronic mail improves my job performance.
4	The electronic mail system addresses my job-related needs.
5	Using electronic mail saves me time.
6	Electronic mail enables me to accomplish tasks more quickly.
7	Electronic mail supports critical aspects of my job.
8	Using electronic mail allows me to accomplish more work than would otherwise be possible.
9	Using electronic mail reduces the time I spend on unproductive activities.
10	Using electronic mail enhances my effectiveness on the job.
11	Using electronic mail improves the quality of the work I do.
12	Using electronic mail increases my productivity.
13	Using electronic mail makes it easier to do my job.
14	Overall, I find the electronic mail system useful in my job.

All things considered, my using electronic mail in my job is:

	Neutral	
Good	:__:__:__:__:__:__:	Bad
Wise	:__:__:__:__:__:__:	Foolish
Favorable	:__:__:__:__:__:__:	Unfavorable
Beneficial	:__:__:__:__:__:__:	Harmful; and
Positive	:__:__:__:__:__:__:	Negative.

2. **Revised six items towards using a chart master:** (Davis, 1989)

Revised 6 items scale for perceived usefulness worded towards CHART-MASTER

Item No.	Candidate item for psychometric measures for perceived usefulness
1	Using CHART-MASTER in my job would enable me to accomplish tasks more quickly.
2	Using CHART-MASTER would improve my job performance.
3	Using CHART-MASTER in my job would increase my productivity.
4	Using CHART-MASTER would enhance my effectiveness on the job.
5	Using CHART-MASTER would make it easier to do my job.
6	I would find CHART_MASTER useful in my job.

Revised 6 items scale for perceived ease of use worded towards CHART-MASTER

Item No.	Candidate item for psychometric measures for perceived usefulness
1	Learning to operate CHART-MASTER would be easy for me.
2	I would find it easy to get CHART-MASTER to do what I want to do.
3	My interaction with CHART-MASTER would be clear and understandable.
4	I would find CHART-MASTER flexible to interact with.
5	It would be easy for me to become skillful at using CHART-MASTER.
6	I would find CHART-MASTER easy to use.

YOUR KNOWLEDGE HAS VALUE

- We will publish your bachelor's and master's thesis, essays and papers

- Your own eBook and book - sold worldwide in all relevant shops

- Earn money with each sale

Upload your text at www.GRIN.com
and publish for free